THE
SPACE
SHUTTLE

PETER MURRAY

THE CHILD'S WORLD

DESIGN
Michael George

PHOTO CREDITS
NASA

Library of Congress Cataloging-in-Publication Data
Murray, Peter, 1952 Sept. 29-
The space shuttle / Peter Murray.
p. cm.
Summary: Discusses the development of reusable
spacecraft and how they have been used and
describes the typical flight of a space shuttle.
ISBN 1-56766-018-5
1. Space shuttles--Juvenile literature. 2. Manned space flight-
-Juvenile literature.   [1. Space shuttles.]   I. Title.
TL795.5.M87   1993   93-7064
629.45'4--dc20

Distributed to schools and libraries in the United States by
ENCYCLOPAEDIA BRITANNICA EDUCATIONAL CORP.
310 South Michigan Avenue
Chicago, Illinois 60604

Imagine that you are on the flight deck of the space shuttle, waiting for liftoff. You are lying on your back, looking straight up through the wraparound windows. You have been waiting, strapped in your seat, for almost two hours. The technicians are checking and rechecking every detail of the launch. The space shuttle has flown many times before, but nothing must be left to chance.

The space shuttle you are strapped in is as long as four school buses parked end to end. It is attached to an enormous, orange fuel tank that is taller than a ten-story building! Connected to the sides of the fuel tank are two solid rocket boosters. The entire shuttle assembly—including the rockets, fuel tank, and fuel—weighs almost five million pounds. Imagine how much power it will take to lift it off the ground and into outer space!

You can hear the countdown through your headset. *Ten…nine…eight.* The main engines start. You hear a loud rumble and feel as if a giant hand has grabbed the shuttle and is shaking it. *Six…five…four.* A huge cloud of smoke billows from the engines. You are still on the launch pad,

but not for long—the solid rocket boosters are about to ignite! *Three…two…one…zero!*

Suddenly it feels as if a charging rhinoceros has hit you from behind. You are pressed back into your seat as the shuttle rises through the cloud of smoke. The shuttle is shaking harder now, bouncing you around in your seat. Within seconds, you are miles above the Atlantic Ocean. The shuttle rotates slowly. Now it seems as if the planet Earth is above you!

Two minutes after launch, the solid rocket boosters have used all their fuel. They fall away from the shuttle. Parachutes open and the rockets drift slowly back to Earth, where they will be recovered for use on future shuttle flights. Your ride suddenly becomes smoother, as the main engines carry you higher. About eight minutes into the flight, the external fuel tank falls away from the shuttle. The giant tank breaks into pieces and burns up as it falls back toward Earth.

It is very quiet now. There is no vibration. You are in orbit around the planet Earth! You are weightless, free to float about the flight deck. Your hair doesn't stay in place, because there is no gravity to hold it down. Looking out the window, you can see Earth. You can see swirls of cloud and the bright blue of the oceans. You can see the shapes of the continents. You are in outer space, traveling at 17,500 miles per hour!

The space shuttle is very different from earlier spacecraft. It is the first space vehicle that is reusable. Suppose that every time you rode

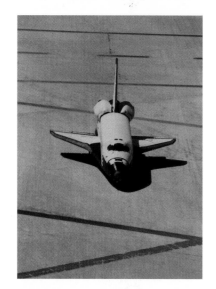

your bike someplace, you had to throw it away and build a new one? That would take a lot of time, and it would be very expensive! But for twenty years, that is how men and women explored outer space. Each spaceship was used only once!

During the 1970s, scientists began working on a new type of space vehicle that could be used over and over. They wanted this new vehicle to take off like a rocket ship, but land like an airplane.

In 1981, the dream became reality. The space shuttle *Columbia*—the world's first reusable spacecraft—took off, orbited Earth thirty-six times, and returned safely to the planet. Since then, four other space shuttles—*Challenger*,

Discovery, Atlantis, and *Endeavor*—have been built. Over fifty successful space-shuttle flights have been launched since 1981.

Inside the space shuttle, you don't need to wear a spacesuit. The flight deck and the living section of the shuttle are pressurized, like the inside of a passenger airplane, so you can wear comfortable, lightweight clothing. The nonpressurized section of the ship, where you cannot go without a spacesuit, is the *payload bay.* It extends from the flight deck all the way back to the engines. As soon as you reach orbit, you must open the payload-bay doors to release the heat that builds up inside the shuttle.

Inside the payload bay is the shuttle's cargo. Corporations and countries pay to have the

shuttle carry their satellites and science experiments into space. That is why the space shuttle is sometimes called the "sky truck." It's like a giant delivery van that goes from Earth to outer space and back again!

On a typical space-shuttle flight, one or more satellites might be placed in orbit. On this flight, you'll lift two satellites out of the cargo bay and place them in orbit using the Remote Manipulator Arm, or "space crane."

Like previous shuttle astronauts, you'll also perform experiments that help scientists plan for the colonization of outer space. In an earlier experiment, astronauts unfolded a sheet of superthin plastic from a tiny package into a solar panel over one hundred feet long. Soon, similar solar panels may harness energy from the Sun for powering space stations and space colonies.

One special satellite, the Long Duration Exposure Facility (LDEF), was put in orbit in 1984 by the space shuttle *Challenger*. The LDEF carried fifty-seven experiments from nine different countries. Later space-shuttle missions returned to the LDEF to check on the experiments. In one experiment, more then twelve million tomato seeds were left in orbit for six years. When the LDEF was retrieved in 1991, the seeds were given to schools and laboratories across the country to see how their time in space had affected them. Nearly all the seeds produced healthy tomato plants!

On some shuttle flights, the entire payload bay is filled with *Spacelab*, a pressurized labora-

tory where scientists work comfortably without spacesuits. Plants, animals, and chemical reactions all behave differently in space because there is no gravity. Spacelab provides scientists with a safe, comfortable place to perform zero-gravity experiments.

On your shuttle mission, you might be asked to go outside the space shuttle to perform experiments or make repairs. Usually, astronauts wear spacesuits and are attached to the shuttle by safety lines—but not always! If you want to really get around in space, you can strap yourself into a Manned Maneuvering Unit. With this "flying armchair," you can be your own spaceship!

Back inside the warm, comfortable living area of the space shuttle, you feel very safe. It is almost like being in a small building, except that there is no gravity. However, these missions to outer space are journeys into the unknown. No one can say for sure what will happen!

The space shuttle is the safest spacecraft ever flown, but on January 28, 1986, the space shuttle *Challenger* exploded seventy-three seconds after takeoff. All seven astronauts died. One of the seven was Christa McAuliffe, a schoolteacher from New Hampshire who was to have been the first teacher in space. The *Challenger* explosion was the worst disaster in the history of the United States space program.

After the *Challenger* disaster, there were no shuttle flights for nearly three years. Scientists checked and rechecked every aspect of the space

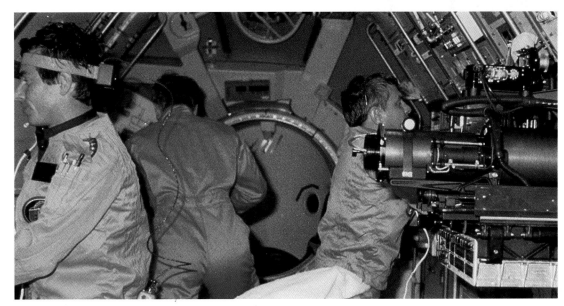

shuttle. They redesigned the solid rocket boosters, replaced the on-board computers, and added an explosive escape hatch. In all, over two hundred changes were made. No one wanted a repeat of the *Challenger* disaster! When the space shuttle *Discovery* finally flew again, everything went perfectly. The space shuttle was back in business!

Since the very first space-shuttle flight, one of the greatest concerns has been bringing the shuttle back to Earth safely. When the shuttle descends through the atmosphere, the friction of the air on the shuttle creates tremendous heat. Without protection, the metal body of the shuttle would melt away. To absorb this deadly heat, scientists insulate the shuttle with a special type of lightweight ceramic tile. The nose and belly of the shuttle are covered with over 31,000 of these tiles!

When it is time to return to Earth, you must slow down by using the shuttle's maneuvering engines. As your speed decreases, the gravity of the Earth draws the shuttle down into the atmosphere. It takes about thirty minutes to glide through the atmosphere to the landing strip in California. When you touch down, you are still traveling at 200 miles per hour. It takes a long time to slow down—it's a good thing the landing strip is almost two miles long!

The future of the space-shuttle program seems secure. More than twenty-five flights have been made since the *Challenger* disaster. The shuttles have successfully launched dozens of

satellites, including the *Hubble Space Telescope*, a satellite that lets us see the farthest reaches of the universe. Soon, the space shuttle will transport materials into outer space to build a permanent space station, where people will live and work. Future space shuttles may even visit the Moon or other planets. Perhaps you will be on board!

INDEX